MEXICO
PACIFIC COAST
BIRDS - AVES

RAINFOREST PUBLICATIONS

Cattle Egret
Garza ganadera
Bubulcus ibis 51 cm

Snowy Egret
Garza nivea
Egretta thula 61 cm

Great Egret
Garza grande
Ardea alba 99 cm

White Ibis
Ibis blanco
Eudocimus albus 64 cm

Reddish Egret
Garza rojiza
Egretta rufescens 76 cm

White-faced Ibis
Ibis cariblanco
Plegadis chihi 58 cm

Roseate Spoonbill
Espátula rosada
Platalea ajaja 81 cm

Black Vulture
Zopilote negro
Coragyps atratus 64 cm

American Kestrel
Cernícalo americano
Falco sparverius 25 cm

Turkey Vulture
Aura cabecirroja
Cathartes aura 76 cm

Bat Falcon
Halcón murcielaguero
Falco rufigularis 30 cm

Osprey
Gavilán pescador
Pandion haliaetus 58 cm

Collared Forest-Falcon
Halcón-selvático collarejo
Micrastur semitorquatus 61 cm

Laughing Falcon
Halcón guaco
Herpetotheres cachinnans 51 cm

Hooked-billed Kite
Milano piquiganchudo
Chondrohierax uncinatus 43 cm

Snail Kite
Milano caracolero
Rostrhamus sociabilis 43 cm

White-tailed Kite
Elanio coliblanco
Elanus leucurus 41 cm

Brown Booby
Piquero moreno
Sula leucogaster 69 cm

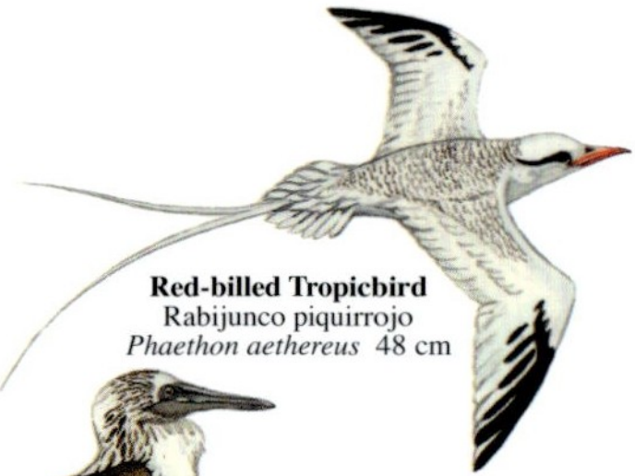

Red-billed Tropicbird
Rabijunco piquirrojo
Phaethon aethereus 48 cm

Blue-footed Booby
Bobo patiazul
Sula nebouxii 81 cm

Anhinga
Anhinga americana
Anhinga anhinga 89 cm

Magnificent Frigatebird
Fragata magnífica
Fregata magnificens 102 cm

Neotropic Cormorant
Cormorán neotropical
Phalacrocorax brasilianus 66 cm

Bare-throated Tiger-Heron
Garza-tigre gorjinuda
Tigrisoma mexicanum 76 cm

Great Blue Heron
Garzón cenizo
Ardea herodias 117 cm

Wood Stork
Cigüeña americana
Mycteria americana 102 cm

Green Heron
Garza verde
Butorides virescens 46 cm

Little Blue Heron
Garza azul
Egretta caerulea 61 cm

Tricolored Heron
Garza tricolor
Egretta tricolor 66 cm

Yellow-crowned Night-Heron
Garza-nocturna coroniclara
Nycticorax violacea 61 cm

Boat-billed Heron
Garza cucharón
Cochlearius cochlearius 51 cm

Black-crowned Night-Heron
Garza-nocturna coroninegra
Nycticorax nycticorax 64 cm

Swainson's Hawk
Aguililla de Swainson
Buteo swainsoni 53 cm

Broad-winged Hawk
Gavilán aludo
Buteo platypterus 41 cm

Common Black-Hawk
Aguililla negra menor
Buteogallus anthracinus 56 cm

Gray Hawk
Aguililla gris
Buteo plagiatus 43 cm

Great Black-Hawk
Aguililla negra mayor
Buteogallus urubitinga 66 cm

Zone-tailed Hawk
Aguililla aura
Buteo albonotatus 51 cm

Short-tailed Hawk
Aguililla colicorta
Buteo brachyurus 41 cm

Red-shouldered Hawk
Aguililla pechirroja
Buteo lineatus 98 cm

Roadside Hawk
Aguililla caminera
Rupornis magnirostris 38 cm

Harris' Hawk
Aguililla de Harris
Parabuteo unicinctus 53 cm

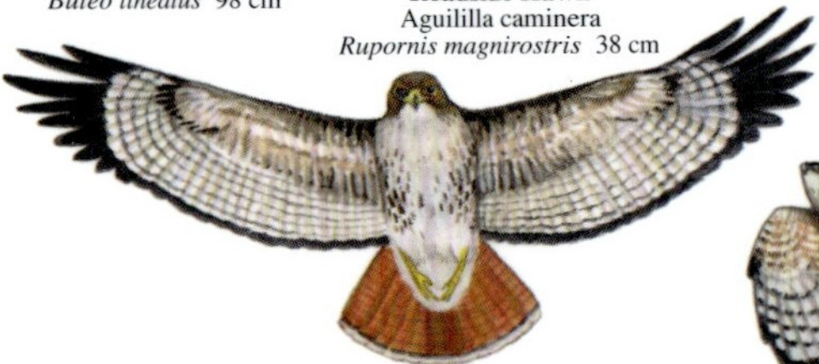

Red-tailed Hawk
Aguililla colirroja
Buteo jamaicensis 56 cm

Cooper's Hawk
Gavilán de Cooper
Accipiter cooperii 38 cm

Crane Hawk
Gavilán zancudo
Geranospiza caerulescens 48 cm

Northern Harrier
Gavilán rastrero
Circus hudsonius 51 cm

Crested Caracara
Caracara común
Caracara cheriway 58 cm

Common Gallinule
Gallineta común
Gallinula galeata 36 cm

American Coot
Gallareta americana
Fulica americana 38 cm

juvenile

Purple Gallinule
Gallineta morada
Porphyrio martinicus 33 cm

juvenile

Northern Jacana
Jacana mesoamericana
Jacana spinosa 23 cm

Spotted Rail
Rascón pinto
Pardirallus maculatus 25 cm

Rufous-necked Wood-Rail
Rascón cuellirrufo
Aramides axillaris 28 cm

Sungrebe
Pájaro cantil
Heliornis fulica 28 cm

Russet-naped Wood-Rail
Rascón cuello gris
Aramides albiventris 38 cm

Marbled Godwit
Picopando canelo
Limosa fedoa 46 cm

American Avocet
Avoceta Americana
Recurvirostra americana 46 cm

Whimbrel
Zarapito trinador
Numenius phaeopus 43 cm

American Oystercatcher
Ostrero americano
Haematopus palliatus 43 cm

Willet
Playero pihuihui
Tringa semipalmata 38 cm

Limpkin
Carao
Aramus guarauna 66 cm

Long-billed Dowitcher
Costurero piquilargo
Limnodromus scolopaceus 28 cm

Black-necked Stilt
Candelero americano
Himantopus mexicanus 38 cm

Lesser Yellowlegs
Patamarilla menor
Tringa flavipes 28 cm

Greater Yellowlegs
Patamarilla mayor
Tringa melanoleuca 36 cm

breeding

Ruddy Turnstone
Vuelvepiedras rojizo
Arenaria interpres 23 cm

Wilson's Plover
Chorlito piquigrueso
Charadrius wilsonia 20 cm

Collared Plover
Chorlito collarejo
Charadrius collaris 15 cm

Semipalmated Plover
Playerito semipalmado
Charadrius semipalmatus 18 cm

Black-bellied Plover
Chorlo gris
Pluvialis squatarola 28 cm

Spotted Sandpiper
Playero alzacolita
Actitis macularius 20 cm

Stilt Sandpiper
Playero zancudo
Calidris himantopus 23 cm

Western Sandpiper
Playerito occidental
Calidris mauri 15 cm

Semipalmated Sandpiper
Correlimos semipalmado
Calidris pusilla 15 cm

Least Sandpiper
Playerito mínimo
Calidris minutilla 15 cm

Laughing Gull
Gaviota reidora
Leucophaeus atricilla 41 cm

Black Skimmer
Rayador americano
Rynchops niger 46 cm

Heermann's Gull
Gaviota de Heermann
Larus heermanni 50 cm

Franklin's Gull
Gaviota de Franklin
Leucophaeus pipixcan 36 cm

Caspian Tern
Golondrina-marina caspica
Hydroprogne caspia 53 cm

Elegant Tern
Golondrina marina elegante
Thalasseus elegans 43 cm

Forster's Tern
Golondrinra-marina de Forster
Sterna forsteri 36 cm

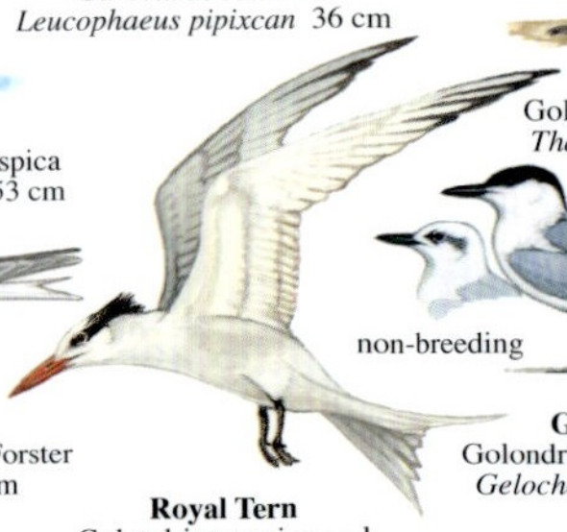

Royal Tern
Golondrina marina real
Thalasseus maximus 51 cm

Gull-billed Tern
Golondrina marina piquigruesa
Gelochelidon nilotica 41 cm

Barn Owl
Lechuza de campanario
Tyto alba 41 cm

Mottled Owl
Búho café
Ciccaba virgata 33 cm

Black-and-white Owl
Búho blanquinegro
Ciccaba nigrolineata 38 cm

Colima Pygmy-Owl
Tecolotito colimense
Glaucidium palmarum 15 cm

Northern Potoo
Bienparado norteño
Nyctibius jamaicensis 38 cm

Common Pauraque
Tapacaminos picuyo
Nyctidromus albicollis 28 cm

Ferruginous Pygmy-Owl
Tecolotito común
Glaucidium brasilianum 15 cm

Lesser Nighthawk
Chotacabras menor
Chordeiles acutipennis 23 cm

Common Ground-Dove
Tórtola común
Columbina passerina 15 cm

Inca Dove
Tórtola colilarga
Columbina inca 20 cm

Ruddy Ground-Dove
Tórtola rojiza
Columbina talpacoti 18 cm

Red-billed Pigeon
Paloma morada
Patagioenas flavirostris 30 cm

White-winged Dove
Paloma aliblanca
Zenaida asiatica 28 cm

White-tipped Dove
Paloma arroyera
Leptotila verreauxi 28 cm

White-fronted Parrot
Loro frentiblanco
Amazona albifrons 25 cm

Mexican Parrolet
Periquito mexicano
Forpus cyanopygius 14 cm

Lilac-crowned Parrot
Loro corona-violeta
Amazona finschi 34 cm

Military Macaw
Guacamaya verde
Ara militaris 76 cm

Orange-fronted Parakeet
Perico frentinaranja
Eupsittula canicularis 23 cm

Broad-billed Hummingbird
Colibrí piquiancho
Cynanthus latirostris 10 cm

Violet-crowned Hummingbird
Colibrí corona-violeta
Amazilia violiceps 11 cm

Ruby-throated Hummingbird
Colibrí gorjirrubí
Archilochus colubris 9 cm

Costa's Hummingbird
Colibrí de Costa
Calypte costae 8 cm

White-eared Hummingbird
Colibrí orfejiblanco
Hylocharis leucotis 10 cm

Mexican Woodnymph
Ninfa mexicana
Thalurania ridgwayi 10 cm

Bumblebee Hummingbird
Zumbador mexicano
Atthis heloisa 7 cm

Calliope Hummingbird
Colibrí de Caliope
Selasphorus calliope 8 cm

Berylline Hummingbird
Colibrí de Berilo
Amazilia beryllina 10 cm

Cinnamon Hummingbird
Colibrí canelo
Amazilia rutila 10 cm

Plain-capped Starthroat
Picolargo coronioscuro
Heliomaster constantii 13 cm

Long-billed Hermit
Ermitaño colilargo
Phaethornis longirostris 15 cm

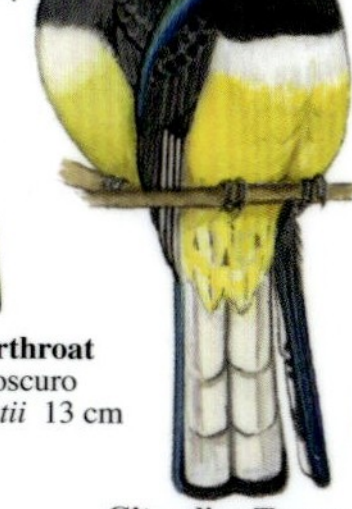

Citreoline Trogon
Trogon citrino
Trogon citreolus 28 cm

Elegant Trogon
Trogón elegante
Trogon elegans 28 cm

Russet-crowned Motmot
Momoto coronicafé
Momotus mexicanus 33 cm

Lesser Roadrunner
Correcaminos menor
Geococcyx velox 45 cm

Greater Roadrunner
Correcaminos mayor
Geococcyx californianus 55 cm

Squirrel Cuckoo
Cuco ardilla
Piaya cayana 46 cm

Groove-billed Ani
Garrapatero pijuy
Crotophaga sulcirostris 30 cm

Lesser Ground-Cuckoo
Cuco-terrestre menor
Morococcyx erythropygus 25 cm

Vaux's Swift
Vencejo de Vaux
Chaetura vauxi 10 cm

Mangrove Cuckoo
Cuco manglero
Coccyzus minor 30 cm

Belted Kingfisher
Martín-pescador norteño
Megaceryle alcyon 33 cm

Green Kingfisher
Martín pescador verde
Chloroceryle americana 18 cm

Olivaceous Woodcreeper
Trepatroncos oliváceo
Sittasomus griseicapillus 15 cm

Ivory-billed Woodcreeper
Trepatroncos piquiclaro
Xiphorhynchus flavigaster 23 cm

Yellow-bellied Sapsucker
Chupasavia vientre-amarillo
Sphyrapicus varius 20 cm

Acorn Woodpecker
Carpintero arlequín
Melanerpes formicivorus 23 cm

Golden-cheeked Woodpecker
Carpintero cachetidorado
Melanerpes chysogenys 21 cm

Gila Woodpecker
Carpintero de Gila
Melanerpes uropygialis 23 cm

Ladder-backed Woodpecker
Carpintero listado
Picoides scalaris 20 cm

Lineated Woodpecker
Carpintero lineado
Dryocopus lineatus 33 cm

Pale-billed Woodpecker
Carpintero piquiclaro
Campephilus guatemalensis 33 cm

Gray-crowned Woodpecker
Carpintero coronigris
Colaptes auricularis 21 cm

Black Phoebe
Mosquero negro
Sayornis nigricans 18 cm

Pacific Slope Flycatcher
Mosquero occidental
Empidonax difficilis 13 cm

Greater Pewee
Pibí mayor
Contopus pertinax 18 cm

Tufted Flycatcher
Mosquero penachudo
Mitrephanes phaeocercus 13 cm

Bright-rumped Attila
Atila rabadilla-brillante
Attila spadiceus 20 cm

Greenish Elaenia
Elaenia verdosa
Myiopagis viridicata 14 cm

Northern Beardless-Tyrannulet
Mosquerito lampiño norteño
Camptostoma imberbe 10 cm

Ash-throated Flycatcher
Copetón gorjicenizo
Myiarchus cinerascens 18 cm

Vermilion Flycatcher
Mosquero cardenal
Pyrocephalus rubinus 13 cm

Brown-crested Flycatcher
Copetón tirano
Myiarchus tyrannulus 20 cm

Dusky-capped Flycatcher
Copetón triste
Myiarchus tuberculifer 15 cm

Sulphur-bellied Flycatcher
Papamoscas vientre-amarillo
Myiodynastes luteiventris 20 cm

Boat-billed Flycatcher
Luis piquigrueso
Megarynchus pitangua 23 cm

Great Kiskadee
Luis grande
Pitangus sulphuratus 23 cm

Social Flycatcher
Luis gregario
Myiozetetes simills 15 cm

Cassin's Kingbird
Tirano de Cassin
Tyrannus vociferans 23 cm

Tropical Kingbird
Tirano tropical
Tyrannus melancholicus 20 cm

Western Kingbird
Tirano occidental
Tyrannus verticalis 20 cm

Thick-billed Kingbird
Tirano piquigrueso
Tyrannus crassirostris 24 cm

Warbling Vireo
Vireo gorjeador
Vireo gilvus 13 cm

♀

Rose-throated Becard
Cabezón degollado
Pachyramphus aglaiae 18 cm

♀

Masked Tityra
Titira enmascarada
Tityra semifasciata 20 cm

Plumbeous Vireo
Vireo plomizo
Vireo plumbeus 14 cm

♀

Black-capped Vireo
Vireo gorrinegro
Vireo atricapilla 11 cm

Yellow-green Viero
Viero cabecgrís
Viero flavovirdis 15 cm

Lesser Greenlet
Verdillo menudo
Pachysylvia decurtatus 10 cm

Mangrove Vireo
Vireo manglero
Vireo pallens 10 cm

Black-throated Magpie-Jay
Urraca-hermosa carinegra
Calocitta colliei 70 cm

Golden Vireo
Vireo dorado
Vireo hypochryseus 13 cm

San Blas Jay
Chara de San Blas
Cyanocorax sanblasianus 32 cm

Purplish-backed Jay
Chara de Beechy
Cyanocorax beecheii 40 cm

Green Jay
Chara verde
Cyanocorax yncas 30 cm

Sinaloa Crow
Cuervo sinaloense
Corvus sinaloae 38 cm

Blue Mockingbird
Mulato azul
Melanotis caerulescens 25 cm

Gray-breasted Martin
Martín pechigris
Progne chalybea 18 cm

Barn Swallow
Golondrina ranchera
Hirundo rustica 13 cm

Northern Mockingbird
Cenzontle norteño
Mimus polyglottos 25 cm

Mangrove Swallow
Golondrina manglera
Tachycineta albilinea 13 cm

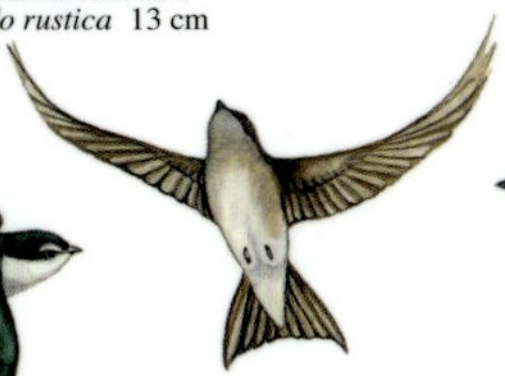

Northern Rough-winged Swallow
Golondrina-aliserrada norteña
Stelgidopteryx serripennis 13 cm

Brown-backed Solitaire
Clarín jilguero
Myadestes occidentalis 21 cm

Happy Wren
Saltapared feliz
Pheugopedius felix 14 cm

Black-capped Gnatcatcher
Perlita Gorrinegra
Polioptila nigriceps 11 cm

Sinaloa Wren
Saltapared sinaloense
Thryophilus sinaloa 14 cm

Blue-gray Gnatcatcher
Perlita grisilla
Polioptila caerulea 11 cm

Gray Silky-Flycatcher
Capulinero gris
Ptiliogonys cinereus 21 cm

House Wren
Saltapared-continental norteño
Troglodytes aedon 10 cm

Rufous-backed Thrush
Zorzal dorisirrufo
Turdus rufopalliatus 24 cm

White-throated Thrush
Zorzal gorgiblanco
Turdus assimilis 23 cm

Orange-billed Nightingale Thrush
Zorzalito piquinaranja
Catharus aurantiirostris 15 cm

Swainson's Thrush
Zorzalito de Swainson
Catharus ustulatus 18 cm

Northern Waterthrush
Chipe-suelero charquero
Parkesia noveboracensis 15 cm

Slate-throated Redstart
Pavito gorjigris
Myioborus miniatus 13 cm

Nashville Warbler
Chipe de Nashville
Vermivora ruficapilla 13 cm

Tropical Parula
Parula tropical
Setophaga pitiayumi 10 cm

American Redstart
Pavito migratorio
Setophaga ruticilla 13 cm

Rufous-capped Warbler
Chipe gorricastaño
Basileuterus rufifrons 13 cm

Fan-tailed Warbler
Chipe roquero
Basileuterus lachrymosa 15 cm

Common Yellowthroat
Antifacito norteño
Geothlypis trichas 13 cm

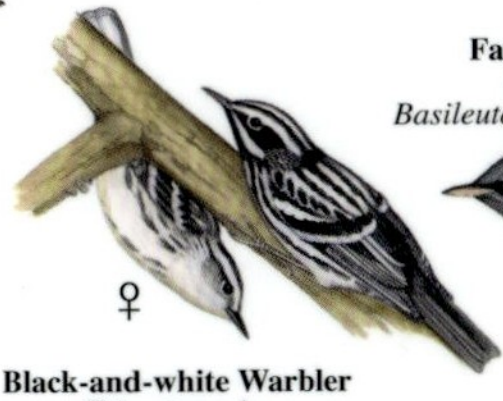

Black-and-white Warbler
Chipe trepador
Mniotilta varia 13 cm

MacGillivray's Warbler
Chipe de Tolmie
Geothlypis tolmiei 13 cm

Wilson's Warbler
Chipe de WIlson
Cardellina pusilla 11 cm

Red-breasted Chat
Granatelo mexicano
Granatellus venustus 16 cm

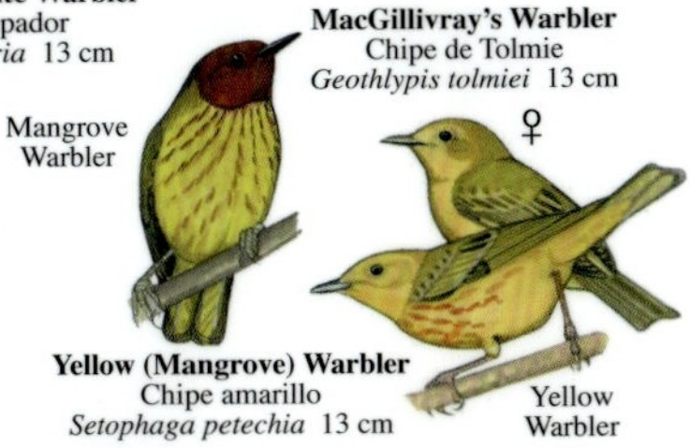

Yellow (Mangrove) Warbler
Chipe amarillo
Setophaga petechia 13 cm

Black-throated Gray Warbler
Chipe negrigris
Setophaga nigrescens 12 cm

Orange-crowned Warbler
Chipe corona-naranja
Vermivora celata 12 cm

Gray-crowned Yellowthroat
Mascarita piquigruesa
Chamaethlypis poliocephala 13 cm

Western Tanager
Tángara occidental
Piranga ludoviciana 18 cm

Red-crowned Ant-Tanager
Tángara-hormiguera coronirroja
Habia rubica 18 cm

Yellow-breasted Chat
Gritón pechiamarillo
Icteria virens 18 cm

Summer Tanager
Tángara roja
Piranga rubra 18 cm

Rosy Thrush-Tanager
Tángara huitlacoche
Rhodinocichla rosea 20 cm

Flame-colored Tanager
Tángara dorsirrayada
Piranga bidentata 18 cm

Ruddy-breasted Seedeater
Semillero pechicanelo
Sporophila minuta 10 cm

White-collared Seedeater
Semillero collarejo
Sporophila torqueola 10 cm

Blue-black Grassquit
Semillero brincador
Volatinia jacarina 10 cm

Rusty-crowned Ground-Sparrow
Rascador coronirrufo
Melozone kieneri 17 cm

Lincoln's Sparrow
Gorrión de Lincoln
Melospiza lincolnii 15 cm

House Sparrow
Gorrión doméstico
Passer domesticus 15 cm

Lark Sparrow
Gorrión arlequín
Chondestes grammacus 15 cm

Chipping Sparrow
Gorrión cejiblanco
Spizella passerina 14 cm

House Finch
Fringílido Mexicano
Carpodacus mexicanus 15 cm

Grayish Saltator
Saltador grisáceo
Saltator coerulescens 20 cm

Striped-headed Sparrow
Zacatonero cabecirrayada
Peucaea ruficauda 18 cm

Blue Grosbeak
Picogrueso azul
Passerina caerulea 18 cm

Red-winged Blackbird
Tordo sargento
Agelaius phoeniceus 23 cm

Blue Bunting
Colorín azulinegro
Cyanocompsa parellina 13 cm

Varied Bunting
Colorín morado
Passerina versicolor 12 cm

Painted Bunting
Colorín sietecolores
Passerina ciris 13 cm

Great-tailed Grackle
Zanate mayor
Quiscalus mexicanus 43 cm

Eastern Meadowlark
Pradero común
Sturnella magna 20 cm

Black-vented Oriole
Bolsero de Wagler
Icterus wagleri 22 cm

Bronzed Cowbird
Vaquero ojirrojo
Molothrus aeneus 20 cm

Streak-backed Oriole
Bolsero dorsirrayado
Icterus pustulatus 20 cm

Hooded Oriole
Bolsero cuculado
Icterus cucullatus 20 cm

Yellow-winged Cacique
Cacique Mexicano
Cacicus melanicterus 33 cm

Orchard Oriole
Bolsero castaño
Icterus spurius 18 cm

Scrub Euphonia
Eufonia gorjinegro
Euphonia affinis 10 cm

Brown Pelican
Pelícano café
Pelecanus occidentalis 119 cm

American White Pelican
Pelícano blanco americano
Pelecanus erythrorhynchos 155 cm

Black-bellied Whistling-Duck
Pijiji aliblanco
Dendrocygna autumnalis 53 cm

Fulvous Whistling-Duck
Pijiji canelo
Dendrocygna bicolor 48 cm

Muscovy Duck
Pato real
Cairina moschata 84 cm

Blue-winged Teal
Cerceta aliazul
Anas discors 38 cm

Least Grebe
Zambullidor menor
Tachybaptus dominicus 23 cm

Pied-billed Grebe
Zambullidor piquipinto
Podilymbus podiceps 33 cm

Cinnamon Teal
Cerceta castaña
Anas cyanoptera 41 cm

Green-winged Teal
Cerceta aliverde
Anas crecca 36 cm

Gadwall
Pato pinto
Anas strepera 52 cm

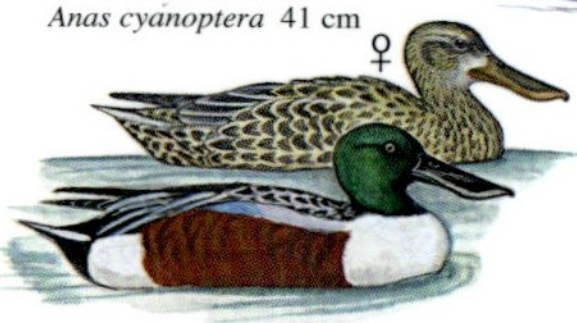

Northern Shoveler
Pato cucharón norteño
Anas clypeata 48 cm

American Wigeon
Pato chalcuán
Anas americana 51 cm

Rufous-bellied Chachalaca
Chachalaca vientre-castaña
Ortalis wagleri 68 cm

Elegant Quail
Codorniz elegante
Callipepla douglasii 24 cm

Illustrated by Robert Dean

Printed in the US on recycled paper. ISBN 1888538-40-6

www.rainforestpublications.com

9 781888 538403